Ingo Schommer

Collaborative
Filtering with RSS

I0009498

Ingo Schommer

Collaborative Filtering with RSS

Research and development of a social feed-reader

VDM Verlag Dr. Müller

Bibliographic information by the German National Library: The German National Library lists this publication at the German National Bibliography; detailed bibliographic information is available on the Internet at http://dnb.d-nb.de.

This works including all its parts is protected by copyright. Any utilization falling outside the narrow scope of the German Copyright Act without prior consent of the publishing house is prohibited and may be subject to prosecution. This applies especially to duplication, translations, microfilming and storage and processing in electronic systems.

Any brand names and product names mentioned in this book are subject to trademark, brand or patent protection and are trademarks or registered trademarks of their respective holders. The use of brand names, product names, common names, trade names, product descriptions etc. even without a particular marking in this works is in no way to be construed to mean that such names may be regarded as unrestricted in respect of trademark and brand protection legislation and could thus be used by anyone.

Copyright © 2007 VDM Verlag Dr. Müller e. K. and licensors
All rights reserved. Saarbrücken 2007
Contact: info@vdm-verlag.de
Cover image: www.photocase.de Julia Nimbus
Publisher: VDM Verlag Dr. Müller e. K., Dudweiler Landstr. 125 a, 66123 Saarbrücken, Germany
Produced by: Lightning Source Inc., La Vergne, Tennessee/USA
        Lightning Source UK Ltd., Milton Keynes, UK

Bibliografische Information der Deutschen Nationalbibliothek: Die Deutsche Nationalbibliothek verzeichnet diese Publikation in der Deutschen Nationalbibliografie; detaillierte bibliografische Daten sind im Internet über http://dnb.d-nb.de abrufbar.

Das Werk ist einschließlich aller seiner Teile urheberrechtlich geschützt. Jede Verwertung außerhalb der engen Grenzen des Urheberrechtsgesetzes ist ohne Zustimmung des Verlages unzulässig und strafbar. Das gilt insbesondere für Vervielfältigungen, Übersetzungen, Mikroverfilmungen und die Einspeicherung und Verarbeitung in elektronischen Systemen.

Alle in diesem Buch genannten Marken und Produktnamen unterliegen warenzeichen-, marken- oder patentrechtlichem Schutz bzw. sind Warenzeichen oder eingetragene Warenzeichen der jeweiligen Inhaber. Die Wiedergabe von Marken, Produktnamen, Gebrauchsnamen, Handelsnamen, Warenbezeichnungen u.s.w. in diesem Werk berechtigt auch ohne besondere Kennzeichnung nicht zu der Annahme, dass solche Namen im Sinne der Warenzeichen- und Markenschutzgesetzgebung als frei zu betrachten wären und daher von jedermann benutzt werden dürften.

Copyright © 2007 VDM Verlag Dr. Müller e. K. und Lizenzgeber
Alle Rechte vorbehalten. Saarbrücken 2007
Kontakt: info@vdm-verlag.de
Coverbild: www.photocase.de Julia Nimbus
Verlag: VDM Verlag Dr. Müller e. K., Dudweiler Landstr. 125 a, 66123 Saarbrücken, Deutschland
Herstellung: Lightning Source Inc., La Vergne, Tennessee/USA
        Lightning Source UK Ltd., Milton Keynes, UK

ISBN: 978-3-8364-2544-5

Thesis by **Ingo Schommer**
Media Production, University of Applied Sciences Darmstadt, 2006

Supervised by Dipl. Media System Designer Franz Spies and Prof. Dr. Arnd Steinmetz

Acknowledgements

This work would not have been possible without the support of the following people - thank you guys!

My first supervisor Franz Spies for motivating me and pushing me to work out a detailed concept before diving into development
My family Theo, Petra, Mona and Karolin Schommer for support throughout my studies - and believing in me
Prof. Dr. Arnd Steinmetz, my second supervisor
Saskia Müller and Simon Brückner for proof-reading
Clemens Gutweiler for his patience in answering technical questions
All alpha-testers at Media Production and elsewere on the net
The open-source-community for inspiration and all those great tools

Table of Contents

Abstract

The increasing amount of information on the internet makes it hard to filter out relevant parts. Every second hundreds of news-articles and blog-entries are written. A subscriber to this constant stream of information quickly looses overview — a typical *needle-in-a-haystack problem*. One couldn't possibly read all of it, but nevertheless has to keep track in case something interesting comes along.

Especially in the field of news-reading, software can help to increase efficiency by personalized filtering mechanisms. This filtering can be predictive based on previous user-choices, but also heeds the power of social networks: Users help each other by marking relevant information.

The goal of this project is the conceptual and practical development of a web-based news-reader with advanced filtering features. Research is conducted in the field of information visualization and filtering as well as in competitive products and their shortcomings. The solution will mainly target pro-users who are already accustomed with news-reading, the modern *information junkies*.

1. Introduction

1.1 Definition "RSS"

RSS ("Really Simple Syndication") provides the technological basis for online sub-scription-services. By automatically generated XML-files, content is distributed in a standardized form to interested readers. The technology quickly gained momentum in the last years: Today it is used by most major news-services and forms a crucial part of every "weblog". Software can be utilized to keep track of updates published via RSS, so called "aggregators" or "RSS-readers" are the digital counterpart of newspaper-collections.

1.2 Motivation and Problem Definition

With the advent of online content syndication, web-users have thousands of news-sources at their fingertips. Applications and web-services compile these sources via RSS and present them instantly. The success of RSS is tightly wrapped with the promise of simplicity, so users naturally strive for simple solutions and less redun-dancy. In a classical newspaper-context, editors ensure that the selected content is interesting to their readers and is not redundant or of low informational value. This intermediate function is not equally present in the new connection between web-publishers and their readers. With time and attention becoming a luxury-item in the world of media, filtering becomes crucial -- a problem that coined the expres-sion of *signal-to-noise ratio*.

The market for RSS-readers is already set by numerous products with a more or less established feature-set. Very much like an e-mail-client, they help users to categorize and sort their content, in this case news-feeds. While this might suffice for a couple of news in the inbox, the concept becomes difficult on a larger scale: RSS-subscrip-tions easily produce several hundred messages each day, exceeding normal email-volumes by far and overwhelming average users. Traditional RSS-readers miss the fact that users demand more help by the application than just putting news into folders.

1.3 Aims and Objectives

The core question is: How do we train dumb computers to be our personal news-editor? The key lies in collaboration with other users who might have the same problem and use the same software to solve it. It works like this: Articles frequently read by other users are also likely to interest myself. This concept tries to use this connection through an additional feature-set, such as a rating-process and news-recommendations by other users.

Take an example: A user has subscribed to both a high-quality feed with only occasional postings, and a popular feed delivering dozens of bulk-entries each day. Chances are that the high-quality postings will be overlooked in the constant stream of messages. By manual rating or a recommendation by a friend, the high-quality news can be highlighted, while others are filtered. News-items have a varying relevancy to each user, and software can help in determining this factor.

Automatically weighted content also saves time: On a busy day, users can block away more information without leaving their RSS-inboxes full of unread news -- the software decides which content is relevant enough to pass the treshold.

The filtering also adds value in different contexts than standard web-surfing: On mobile devices, the amount of information is often limited by screen size and bandwidth -- a scenario where filtering is just an economic necessity.

1.4 Roadmap

Based on the prerequisites by the current examination regulations in Media Production, the roadmap is split into a *Research/Concept-Phase* ending with the winter-term 2005/06 and an *Implementation-Phase* due at 2006-06-19.

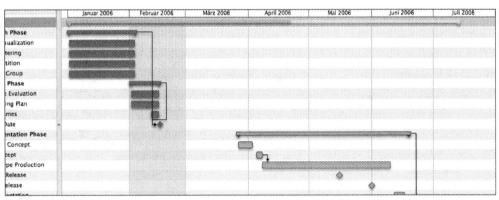

fig. 1-1: Roadmap

2. Research

2.1 Information Visualization Methods

The amount of information that needs to be processed in the modern world leads to an expansion and re-interpretation of traditional methods to visualize information. Evolution of alternative methods is also propelled by the possibilities of electronic devices and their interactivity. This chapter deals with the unique ways that information can be reproduced in graphics and structure, with a close look on its feasibility in RSS-reading.

Linear

Probably the easiest and most popular method of structuring is an one-way linearized representation with optional separators and sorting – the common choice in analog media such as books and newspapers. This form corresponds to the very nature of text itself, and therefore is also the preferred method on content-oriented websites – with several useful additions such as bookmarks and hyperlinks.

In terms of a news-aggregator, this behavior is reflected by the "river-of-news"-view – all unread news are displayed on one long page, separated by the originating feeds. The analogy goes back to the inventor of RSS, Dave Winer: "*(...) there's another kind of reader, an aggregator, that works differently, and I think more efficiently for the human reader. Instead of having to hunt for new stories by clicking on the titles of feeds, you just view the page of new stuff and scroll through it. It's like sitting on the bank of a river, watching the boats go by. If you miss one, no big deal. You can even make the river flow backward by moving the scollbar up.*"[1]

Hierarchies

Complex information needs to be separated into smaller logical units: They are linked either vertically or horizontally, with a classical parent-child-relationship as well as sibling-linkage. Grouping on multiple levels enables the recipient to get a quick overview on a large amount of information. He might hide subordinate layers to reduce the amount of information, or unfold additional groups for a more detailed view. In computer-terms this equals sitemaps of web-pages and the structure of file-systems ("directories" in the desktop-metaphor),

Most applications that have a distinct group-entity-relation in their datasets use hierarchies. The classical example is an e-mail-client such as Microsoft Outlook, separating the data into three different views: folders, e-mail-titles and content. The majority of news-aggregators treat news-items like e-mails and adopt the three-column-view.

---

1 Winer, Dave. "What is a 'River of News' style aggregator". 10 Feb. 2005 <http://www.reallysimplesyndication.com/riverOfNews>.

Graphs and Diagrams

Originating in mathematics, a graph is used to describe numerical data-sets measured along a set of two- or three-dimensional axis.

A special form of graph is the "scatterplot", which additionally assigns weighting to the single coordinate-points as a point-cloud, introducing a further dimension for statistical interpretation (conf. *fig. 2-4*).

Graphs mainly visualize relationships and trends in large amounts of data, but are difficult for displaying data with multiple relationships and for structuring text. As news-items barely contain any numerical data, this form of representation is unsuitable.

Diagrams on the other hand are also structured representations, but not bound to distinct axis – they can take various forms: Mind maps to interconnect ideas in a twodimensional space, Gantt charts for project management, UML to visualize programming structures, flowcharts to display processes etc.

Networks

A network is used to connect specific items by common attributes. Each *node* can have multiple connections and is not neccessarily bound to a hierarchy of any kind. Networks work best where equal items are closely interrelated. Each item in the network can be compared to another item to detect similarities and a common "neighborhood".

Most collaborative tools base on networking, e.g. connecting people to a social network based in their hobbies or shared friends *(conf. fig. 2-3 Vizster)*.

Networking can also provide an alternative view on text, e.g. the project *Textarc* (conf. *fig. 2-7 Textarc*) which displays a whole text on one screen page in an "*funny combination of an index, concordance and summary*"[2]. Words are presented clockwise in a graphical arc, drawn closest to where they appear in the text and lighter if they are used frequently.

Tags

This technique is not a new form of data-representation, but nevertheless a very popular method on the internet – a solid basis for collaboration. *Tags* are meta-information that is related to a piece of information and categorize it. They have developed out of the insight that different people might group data-sets by varying keywords – a fixed categorization evident in directory-services such as Yahoo.com or Dmoz.org wasn't practical in collaborative software.

One of the most popular uses is "del.icio.us", a *social bookmarking system* where bookmarks are related to user-defined tags. This enables searching for websites based on their semantic significance, e.g. a product-review on the new iPod Nano might be found by the tags "iPod", "MP3-Player" or "Apple".

---

2 *Bradford Paley, W. "Textarc.org". 10 Feb. 2006 <http://www.textarc.org>*

fig. 2-1: newsmap
Source: marumushi.com/apps/newsmap

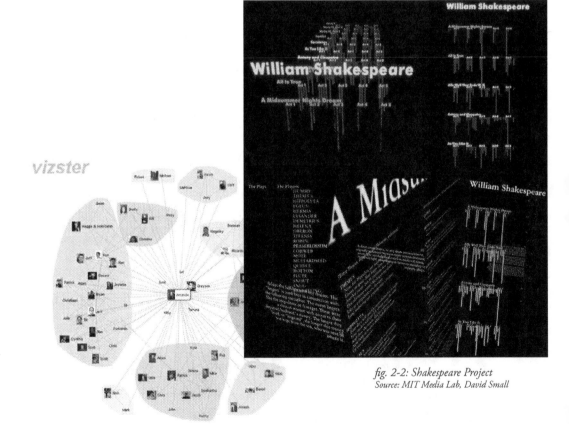

fig. 2-2: Shakespeare Project
Source: MIT Media Lab, David Small

fig. 2-3: Vizster, visualizing online social networks
Source: jheer.org/vizster

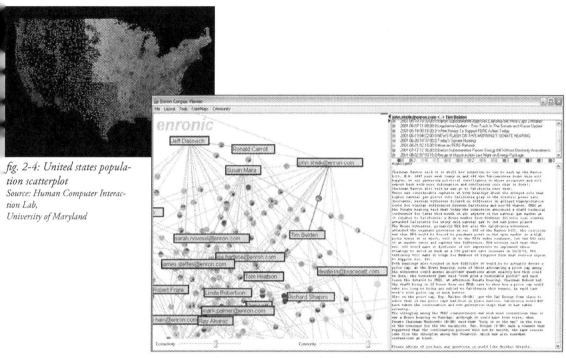

fig. 2-4: United states population scatterplot
Source: Human Computer Interaction Lab,
University of Maryland

fig. 2-5: Enronic, networked view of company-wide email-communication
Source: jheer.org/enron

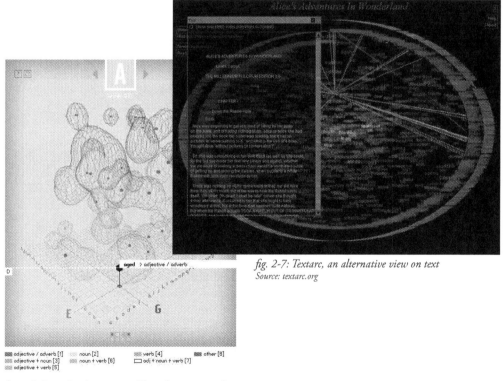

fig. 2-7: Textarc, an alternative view on text
Source: textarc.org

▨ adjective / adverb [1]   ▨ noun [2]   ▨ verb [4]   ▨ other [8]
▨ adjective + noun [3]   ▨ noun + verb [6]   ☐ adj + noun + verb [7]
▨ adjective + verb [5]

fig. 2-6: Base26, 3d-mapping of four-character-words
Source: Human Computer Interaction Lab, University of Maryland

In the news-context *tagging* is already changing the ways existing feeds are structured by communities and new ones can be discovered. Little effort has yet been made to tag specific news-items.

Treemaps

So called *Treemaps* show hierarchical data in a space-constrained area by different rectangles which each contain another set of children-rectangles. Larger rectangles signify a larger amount of data – or simply more relevance. Unlike vast tree-node representations, this recursive approach provides a quick overview on the given data-set – and enables users to "zoom into" an area for further detail.

Web-applications have used this approach to visually order a set of items by relevancy, e.g. the most popular news-items on Google News, where less important news only get a minimum of screen "real estate" while headliners are prominently placed (conf. *fig. 2-1 Newsmap*). The downside here: Titles may be cropped or scaled beyond readability – a treemap is not easily scaleable to hold longer textual information.

Three-dimensional

The ability to project data in more than two dimensions is mentioned here for sake of completeness rather than posing a real alternative for displaying news-items. Most of the previously mentioned methods can be potentially transformed into a 3d-visualization: Data-items could be sorted by transparency and their depth in space, while the user can "walk" through the data.

As a logical progression of interlinking documents via *hypertext*, entire libraries can be visualized in the third dimension, e.g. the complete work of Shakespeare (*conf. fig. 2-2 Shakespeare Project*): Books are ordered into one dimension, with the connected act-structure into another axis. Textual information is viewed at different scales, from a quick overlook to single pages and cross-links. Various visual filters can be used to structure the texts.

Conclusion

The usage of the presented methods heavily depends on the data that needs to be displayed, in the case of a news-aggregator mostly date-based chunks of text and image with attached meta-information. A visualization of this chunks in graphs or networks would not work, because the items are barely interconnected. Threedimensional display is possible in theory, but in the end a RSS-reader is a just tool that has to work according to users expectations: Due to lacking experience and incapable input-methods, the way into the third dimension is difficult for simple text-representation. The generally low adoption-rate of threedimensional visualization is evident both from the presented examples and the market overview on existing RSS-readers (conf. *fig. 2.3 Competitive Analysis*). The proposed application will use a combination of hierarchical view on feeds and a "river-of-news"-style on news-items to form an efficient solution.

2.2 Information Filtering Methods

Filtering of information works on many levels — every person naturally filters their sensual impressions to get an understandable amount of relevant data. On the internet, where information is abundant, we need even more filtering – based on our specific interests and behavior, a process called *personalization*. Personalization systems can be classified into three types: rule-based systems, content-based filtering systems and collaborative filtering systems. Established methods of all three classifications are evaluated in respect of usefulness for news-reading.

Sorting

Sorting is the process of arranging information according to a specific key. The resulting order might be alphabetic, numerical — or in a news-context mostly date-based. While sorting does not directly mean *filtering*, it comes in handy to prioritize content, putting the least relevant information to the bottom of a page, or hide it by putting a limit on the displayed entries. The concept of *charts* in music and cinema basically works the same way.

Categorization

Categories are equal with the concept of *groups* and *folders* in a RSS-reader. They attach semantics to an item by putting it in relation with other items. Categories can be flat or nested, depending on the current need. In the news-context, *feeds* are the most basic grouping of news-items. These feeds are again grouped into logical units, e.g. "Sports" or "Family".

User-based Recommendations

Users often feel compelled to publish their personal opinions on products (or news items) — think Amazon User Comments. This manual user-intervention is a solid basis for information filtering, as it directly relates to a product and reflects a personal opinion. Nevertheless, opinions tend to be expressed if a product is either very good or overly bad, so the filter-results are not evenly spread.

A variation of user-based recommendations are *editorial* recommendation: A group of editors selects items they deem appropriate for a specific group of users. This model is popular on linkblogs such as Slashdot.org or Digg.com, which are basically compiling special-interest links around the net and present them with a short description.

Neighborhood Similarities

The *K-nearest neighbor algorithm* tries to predict correlations, either on a user-to-user basis ("These users also liked X") or item-to-item basis ("Users who bought X also liked Y"). It examines a set of features for similarities, and computes neighbor-datasets based on *distance-vectors*. This algorithm needs a large number of "trained" records to function accurately, and requires a great deal of computing as it has to examine each available node to achieve full accuracy. For scalability reasons, data-sets are often *clustered* in smaller units.

A short example: User A and user B both liked the movie "Kill Bill", while user B and user C liked "From Dusk till Dawn" as well. This leads to the conclusion that user A might like "From Dusk till Dawn" as well – the item is in his neighbour-hood. If a fourth user D is included, which likes "From Dusk till Dawn" and "Reservoir Dogs", the latter movie is indirectly related to user A with a distance vector of 2.

fig. 2-8: Amazon Recommendations
Source: *amazon.com*

Empirical Predictions

An application can track user-behaviour and treat it as demographical data, e.g. how much of the news-items in a feed or group are actually read. In some situations, a system can learn from recorded behavior, and predict future actions. A RSS-reader might prioritize a feed where occasional postings are thoroughly read by the user above a high-volume feed where only a small percentage of news is actually read. This predictions can also be expanded into a collaborative effort, where actions of different users are compared and averaged.

Adaptive Filtering (Bayesian Method)

Bayesian filtering essentially is a special form of empirical prediction. It became popular with the email spam-problem. In this context, the Bayes theorem states that words in an email which is marked as spam have a certain probability to occur again in future spam-mails, divided by the probability that the word might occur in any (possibly non-spam) email[3]. Each word in an email contributes to the spam-probability — if this factor reaches a certain treshold, the email is marked as spam. By this means, an adequately trained algorithm can achieve hit rates of over 95% without manual user-intervention. The power of Bayesian filtering are its efficiency with large amounts of text as well as the fact that it personalized: each user can train the algorithm differently.

---

3 Graham, Paul. "A Plan for Spam" Aug. 2002 <http://www.paulgraham.com/spam.html>.

Bayesian filtering can also be transferred to news-reading, where the computed hit rates contribute to the overall relevancy of a single item. A user marks news as uninteresting "spam" by just not clicking on them. Nevertheless, this method might not work as expected in every case: while several keywords might be a good indicator for relevancy, the "bad keywords" are not as distinctive as in spam-messages.

Conclusion

Information-filtering in news-reading is a complicated process that can't be scientifically boiled down to simple formulas - even the most elaborate filters such as *Amazon Recommendations* sometimes provide items which do not match a users profile. The variety of presented methods indicates that *personal interest* can be very diverse and hard to compute. Like in most scenarios, the right mixture of methods is the key to success. Factors such as failure-rate, personalization-possibilities, training-effort, the grade of manual user-intervention and scalability each play their roles in choosing the right mix.

2.3 Competitive Analysis

In parallel to the gained popularity of RSS during the last three years, several tools emerged into the new market, with big players such as *Google, Yahoo, AOL* and even *Microsoft* jumping into the field. *Apple's Safari* and *Mozilla Firefox* already have built-in support of RSS, with limited reader-capabilities. The upcoming *Windows Vista* will include additional functionality into Internet Explorer, bringing RSS to the masses.

Some of the applications are developed for users new to RSS with a limited and easy-to-use feature-set. Others are targeted towards "pro-users". The attached competitive analysis is focused on both, with an eye on features and first attempts of collaboration-features. It might not represent a balanced view on the market of RSS-readers – innovative newcomers are favored over established products to present their interesting features.

Google News (news.google.com)

Google News tracks different standard news-sources, and groups news-items by relevance. Users are enabled to personalize their news-pages with country-specific news and special-interest topics. The service does not provide tracking of custom RSS-feeds other than the selected standard-set, so it is not a "RSS-Reader" by definition. By restricting its categories to a handful of "main-interest"-regions and putting focus on daily news, the service reaches a large user-base, but won't suffice for the more demanding users. Google News is nevertheless an interesting competition mainly due to one unique feature: It groups and filters news-items solely via algorithms – human intervention is not needed and in fact forbidden in regards of political viewpoints and subjective selection-criterias.

Newsgator (newsgator.com)

As a "rising star" in RSS-business, Newsgator provides everything from software-clients over web-based readers to company-wide application servers for productivity-solutions and internal information-management. To ensure interoperability, all these services need to be synced – which isn't necessary for a web-only approach. It features a rating-system where users can vote on each news-item. If a item was previously voted by other users, the rating is displayed. The overall relevance of the feed is computed with this voting-process. Unfortunately the information isn't leveraged for filtering-purposes.

Bloglines (bloglines.com)

Bloglines is the "jack-of-all-trades" in web based RSS-software: With folder-based subscriptions it eases the management of feeds greatly, a mobile version allows display on cellphones, custom notifiers are available. It even includes a custom weblog-system, where users can blog their newest feed-discoveries. An interesting feature is the ability to "share" selected parts of the personal feed-list online – a first step into collaboration.

Listmania!
Fabulous Filmmakers: Quentin Tarantino
Tom Seman (Springfield, VT USA)
Qualifications: Reviewer etc

Reservoir Dogs : (Mr. Brown) 10th Anniversary Special Limited Edition
DVD ~ Kirk Baltz
Used & New from: $7.70
Tarantino's fantastic debut, a remake of a 1987 Hong Kong film, "City on Fire". Very impressively directed and written.

True Romance (Unrated Director's Cut) (Two-Disc Special Edition) DVD ~
Christian Slater
$13.47   Used & New from: $11.97
Tarantino wrote the screenplay for this film to finance "Reservoir Dogs".

Natural Born Killers DVD ~ Woody Harrelson
$12.99   Used & New from: $6.47
Tarantino wrote the first draft of the script, but it was almost entirely re-written.

Pulp Fiction (Collector's Edition) DVD ~ Rosanna Arquette
$12.99   Used & New from: $9.97
Without a doubt Tarantino's finest work. Flawless.

fig. 2-10: User-compiled lists on Amazon
Source: amazon.com

Top Tracks

For the week prior to Monday January 30, 2006

Position	Last	Change	Track	Plays	Reach
1	1	non-mover	The Postal Service - Such Great	16757	
2	2	non-mover	Death Cab for Cutie - Soul Meets Body	16235	
3	3	non-mover	Coldplay - Fix You	12058	
4	100	up 96	Arctic Monkeys - I Bet You Look Good	17646	
5	8	up 3	Oasis - Wonderwall	10367	
6	6	non-mover	The Arcade Fire - Rebellion (Lies)	10153	
7	5	down 2	Coldplay - Speed of Sound	9439	
8	7	down 1	Bloc Party - Banquet	10222	
9	4	down 5	The Strokes - Juicebox	13426	
10	108	up 98	Arctic Monkeys - Fake Tales of San	14473	

fig. 2-11: Last.fm Charts, rated by total play-count
Source: last.fm

Findory

Signal vs. Noise

fig. 2-12: Feed-neighbourhoods in a tag-cloud
Source: findory.com

Recently Tagged Stories

- "Smart" Sharpening with The GIMP 🔲 gimp
- RadioTux Sendung 02/2006 🔲 februar
- Feature: The inbox makeover 🔲 email
- RDF schema for citations 🔲 rdf
- Production and Locations for Dillard's 2005 Holiday Book 🔲 locations.cout
- Feature: The inbox makeover 🔲 mac
- Online Tools 🔲 web-based
- Sunni tribes turn against jihadis 🔲 read 1
- ROJO: EXCELENTE PARA LEER NOTICIAS 🔲 noticias
- Google, the Naked Emperor 🔲 google

Frequent Tags

accessibility adsense advertising ajax
apple art attention audio baseball bbc
blogging bloglines blogs bluetooth b
broadband browser bush business ci
cool copyright crime CSS del.icio.us
development digg diy download drm dvd
economics education email energy enviro

fig. 2-15: Story-Tagging by Rojo
Source: rojo.com

⟩ Reuters: Technology

21,055 subscribers | unsubscribe

Goldman says Internet is back, but with provisos

ATHENS (Reuters) - The market for internet companies to go public is back, the managing director of Goldman Sachs' GS.N high-tech group said on Sunday, but only with provisos and with private equity firms increasingly calling the shots.

10/16/2005 11:45:43 AM ◇◇◇◇◇
⊞ 9 incoming links

Court enters final judgment in Lexar Toshiba case

SAN FRANCISCO (Reuters) - Lexar Media Inc. LEXR.O said on Friday that a California court entered a final judgment in the memory card maker's case against Toshiba Corp. 6502.T in which Toshiba must pay the more than $465 million.

10/14/2005 6:15:03 PM ★★★★★

HP says recalls 135,000 notebook PC batteries

SAN FRANCISCO (Reuters) - Hewlett-Packard Co. HPQ.N has recalled 135,000 battery packs for its laptop computers, citing a fire hazard, the world's second-largest personal computer maker said on Friday.

10/14/2005 4:24:55 PM ★★◇◇◇

fig. 2-16: "River-of-news" with rating at Newsgator
Source: newsgator.com

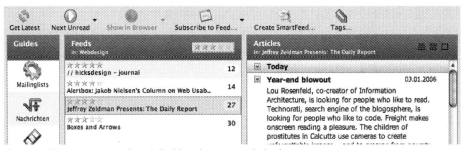

fig. 2-17: Three-pane approach with feed-based rating on Blogbridge
Source: blogbridge.com

Rojo (rojo.com)

Unlike most RSS-readers, Rojo strongly relies on community-features: It accumulated stories and feeds into "most-read"-lists and lets people have a social network based "shared-lists". The use of "tagging" is a key-factor to the success of community-driven sites and fits nicely into the concept of collaborative RSS-reading: Aside from structuring personal feeds into folders, users are enabled to describe information using short descriptive names, so called "tags". This meta-information is used for primitive filtering, in case of Rojo into "Stories tagged by me", "Stories tagged by my contacts" and "Stories tagged by everyone". Unfortunately this approach usually brings in more information rather than filtering the existing feeds – unless you totally revoke the concept of having a predefined set of subscriptions and start reading tags instead of feeds.

Blogbridge (blogbridge.com)

The only desktop-application in this analysis is based on Java, making it available for many platforms. Blogbridge features a common three-pane interface, with the usual grouping- and folding-mechanisms. Aside of that, it is probably the most exciting candidate in the competition, due to several unique features:

Users can set tags on each feed, and share these tags with the community by a semi-automatic upload-process. New users can leverage this feature by getting feed-recommendations.

Another significant feature of Blogbridge is called *Blogstarz*, a feed-based rating-system. The application automatically suggests a rating based on other users, but can be overwritten with a custom rating from one through five. This enables a basic filtering, e.g. by excluding feeds that are rated below three. Unfortunately, the rating-system does not expand to specific news-items.

Pro-users will probably be interested in the *Smartfeeds*-option: Much like *Smart Folders* on *Mac OS X Tiger*, a continuous search on predefined keywords and various other conditions presents the user with results. In combination with ratings and the possibility to limit queries on the subscribed feeds, Smartfeeds pose an interesting alternative approach on feed-reading.

Others

A special blend of RSS-readers are e-mail-clients with integrated RSS-functionality, such as Mozilla Thunderbird. While this approach might succeed with a limited amount of news, similar to newsletters and mailing-lists, it fails as a full-fledged reader-application: The constant news-stream adds too much noise to more important email-messages, which could be counterproductive. More exotic approaches also include console-based readers for the die-hard-users and various browser-plugins.

The competitive research included several other products that are not directly introduced because most of them overlap in their feature-set:

AOL Feeds (feeds.my.aol.com)
Feedlounge (feedlounge.com)
Findory (findory.com)
Pluck RSS Reader (pluck.com/products)
Raggle (raggle.org)
Rocketinfo (rocketinfo.com)
RSSor (rssor.utblog.com)

3. Concept

3.1 Marketing

Targeted Users

The average user of the proposed system is technically aware and has most likely used RSS before. He is interested in a broad range of different fields, and needs to keep track of current developments instantly. He might already use a RSS-reader to structure his large amount of subscriptions, but is always open to new innovations and willing to switch if a better solution comes along. His profile would be best described as a *power-user* with several hundred news-items per day. Aside from RSS-reading he manages large clusters of information with modern technology, such as bookmarks and todo-lists. Most of the time, he has access to the internet and isn't afraid to use web-based tools. He might have experiences in collaborative approaches such as wikis, social-bookmarking and friend-networks.

Licensing Model

The application will profit greatly from open-source-software such as PHP, MySQL, Eclipse, Unix, Linux and several open-sourced code-libraries, but also from the support and knowledge acquired on the internet free of charge. Given these facts, the decision to put the application under an open-source license was easy. The harder part was actually choosing between the numerous licensing models: GPL, LGPL, Mozilla Public License, BSD License, Apache License, to mention but a few.

The decision fell for the popular *GNU General Public License v2* [4], mainly because open-source libraries that are planned to be used in the application require a GPL-compatible license. The GPL uses the concept of *copyleft*, whereas all software can be copied, redistributed and modified as long as it stays under a GPL-compatible license.

Open-source licensing also enables access to SourceForge (*www.sourceforge.net*), a software development platform with free storage, bugtracking and version-control management. In regards of future development, it also eases collaboration with multiple developers. In accordance with the GPL, the application's source will be available on SourceForge.

Promoting Social Software

Due to the non-commercial nature of the product, it has to be promoted without any budget. This is not a big obstacle, if applied correctly: Good online marketing is mostly viral, engaging the user as a marketer. Good software will spread via word-of-mouth, through blogs, linklists, forums and instant messaging. In some

---

4 *GNU Project. June 1991 <http://www.gnu.org/licenses/gpl.html>*

cases, a single link on popular sites such as Slashdot (*www.slashdot.org*) or Digg (*www.digg.com*) can boost user-amounts exponentially. If users believe that the software has value for them, they will promote it to other users – a process called *evangelizing*. A short example: The startup-company *Skype* (*www.skype.com*) based on a simple idea: providing a software for making phonecalls through the internet, free of charge. In only some years, the software has more than 40 million users worldwide[5], where most of their early customer-base was acquired by the infamous word-of-mouth.

The only thing a developer has to ensure is that the software actually *is* valuable in some way. New software-concepts generally attract a lot of *early adopters*, internet-savy people with an open attitude towards innovations. It is easy to grab a first look from early adopters, but if the software doesn't live up to the initial expectations they will just a quickly turn away. Their attention-span is just too short to deal with flawed or uninteresting software. It is therefore crucial to release thoroughly tested versions, even if they bear an alpha-label – and leave features which are not fully working for a later release.

Preview-websites are also an effective promotional mean: Without showing too much of the actual product, they briefly outline the planned features and spare screenshots. They mostly include a form where potential users can subscribe to further notifications via email – a loose customer-binding that is valuable for both sides.

Developers themselves can use blogging as a promotional effort: By maintaining a vivid developer-blog, they stay in touch with potential users – and bring their product to the user's attention by frequent posts and updates.

Financing

Although most open-source software is provided free of charge, this does not imply that there are no costs in producing and sustaining such a product. The business-models of open-source are various: Some licensing models such as the GPL allow users to actually sell their software as long as it's sourcecode is attached[6]. Other open-source developers supply themselves by selling documentation and support for their product. Again others are using venture capital or have a supporting company (e.g. the *Eclipse Project* is supported by *IBM*).

None of these models will work out for this project on it's current scale. RSS-readers are already available for free on the web, and users won't be willing to pay for a new solution unless it provides significant advantages.

The product will be produced without any costs, as development-time is contributed freely by the author. Other developers will hopefully join the project with the same spirit of open-source, helping in support, bug-fixing and further development. Promotion and distribution are not necessary in a traditional sense, so a budget won't be needed here.

---

5 *BusinessWeek online. "The power of Us". 20 Jun. 2005 <http://www.businessweek.com/magazine/ content/05_25/b3938601.htm>*
6 *GNU Project. "Frequently Asked Questions about the GNU GPL". 19 Aug. 2005 <http://www.gnu. org/licenses/gpl-faq.html#DoesTheGPLAllowMoney>*

The only econonomic factor that needs to be considered is server- and bandwidth-ressources: The application involves some serious calculation of rating-averages, caching and parsing – a dedicated server-platform will be mandatory. In terms of bandwidth-usage, the application will also take it's toll by reloading several thousand feeds in each update-process and delivering the results to the user. The ressource-limits of a dedicated server are conservatively estimated to a couple of hundred users. Both server- and bandwidth-costs are generally low on the German market, so the monthly costs will be sustained by the developers in the initial phase.

If the application gets overly popular, the business model will have to be reconsidered. Algorithm-efficiency has to be evaluated, as well as the frequency of updating-runs. Simply asking established users for a tiny contribution often helped open-source-projects to overcome these difficulties – the development-platform *SourceForge* already provides a donation-process for each registered project. Another popular approach is to provide advanced functionality for minimum monthly costs, e.g. more frequent updating or more elaborate algorithms.

3.2 Platform Considerations

The decision was basically between a desktop-based and a web-based application. Both approaches have their pros and cons, and successful RSS-readers can be found on both sides. It is a question of features as well as personal preference.

A desktop-application has the advantage of fast data-processing and a responsive interface that the user is accustomed to. News-feeds are cached locally, making them available even when the user is offline. If the same feeds are accessed on a different machine, desktop-applications might not determine the current "read-status" of a news-item – or have to synchronize first. The tight integration into an operating-system makes desktop-software dependent on a platform – to reach all potential customers, the software has to be developed in parallel for multiple platforms with a much higher workload. Java-based products work around the platform-dependency-problem, but tend to bear an unresponsive interface that is not fully integrated into the targeted operating system. Desktop-newsreaders generally include a stripped-down version of a browser for viewing HTML-formatted content. Many beginning users value system-notifications (sounds, unread-count) when new content is received – a feature that is only available to desktop-software. Nevertheless that specific feature doesn't fly for power-users – just because there is *always* new content.

On the other side, a web-application takes advantage of an application that already exists on every computer: the web-browser, an environment that every user is accustomed with. Many features are already available: History, bookmarks, fulltext-search, text-resizing, et al.. Modern web-applications can emulate the behavior of desktop-software, with the advantage of being platform-independent and relatively easy to maintain. One downside is the requirement of a permanent internet-connection – although that shouldn't be a problem in times of ADSL-flatrates and the targeted group of power-users. The process of RSS-reading itself is often tightly bound to an online-connection: Even if content is locally cached, a user might want to read the full article behind the feed-excerpt, follow posted links, view comments on an article, stream a video-file or enlarge an included thumbnail – actions that would require the user to leave a desktop-application and switch to a browser. In an "always-on" web-based solution, retrieving external data just requires a new tab in the already open browser-window. Power-users often emulate the missing caching of web-based applications by just opening an additional tab for each article for later reading.

3.3 Features

Due to the short time frame of the project, the software probably won't leave the beta-state, but that shouldn't concern users: Google's *Gmail*-service is in public beta for over 18 months now, which had no effect on it's popularity. The roadmap is to have a basic feature-set needed for RSS-reading available for an *alpha-phase*, which probably accounts for a large amount of the total development. The attached *beta-phase* will then incorporate the advertised key-features as a proof-of-concept. Usage and response of beta-testers will be evaluated, followed by a *public beta* as the preliminary result of this project. The projected features will be prioritized and categorized according to this roadmap.

Template-based Output

Modern CSS- and XSL-based templating will help to separate the presentation from the data, and provide a customizable interface to the user.
Relevancy: ***** *Implementation: Alpha Phase*

RSS/Atom-Viewing

Established standards such as RSS 0.91[7], RSS 2.0[8] and Atom[9] will be supported, with images and evaluation of the attached meta-data.
Relevancy: ***** *Implementation: Alpha Phase*

User-Management

To provide a personalized access, users will identify themselves via a user/password-combination. Because the need for privacy and data-protection is relatively low, the possibility to store authentication as a permanent cookie will be available. By this method, users will be automatically logged into their account even after restarting their browser or system. A user is able to change his personal data and application-settings on a brief preference-screen.
Relevancy: ***** *Implementation: Alpha Phase*

7 *Libby, Dan. 10 Jul. 1999 <http://my.netscape.com/publish/formats/rss-spec-0.91.html>.*
8 *Really Simple Syndication: RSS 2.0.1 Specification (revision 6). 25 Jan. 2006 <http://www.rssboard. org/rss-specification>.*
9 *The Atom Syndication Format. Dec.2005 <http://www.ietf.org/rfc/rfc4287.txt>.*

Bug-Reporting

Being a crucial part of software-development, a small bug-tracker can help to organize the bugfixing-workflow.
*Relevancy: ***** Implementation: Alpha Phase*

OPML-Import/Export

The *Outline Processor Markup Language*[10] is a quasi-standard in storing groups of RSS-feeds in an XML-format. Nearly every RSS-reader can import from and export to this format, making it a crucial feature for switchers: A power-user probably won't enter dozens of feed-adresses manually into our new application.
*Relevancy: ***** Implementation: Alpha Phase*

River-of-News-View

By clicking on a folder or feed, a second panel displays the actual news-content, automatically marking the content as "read". News-items can be expanded and collapsed, as well as manually marked as "unread".
*Relevancy: ***** Implementation: Alpha Phase*

Automatic Updates and Caching

The feed-updating process is either activated on user-request or in short intervals (approx. 10 minutes). To maintain a reasonable performance, fetched feed-content must be cached on the server until the user retrieves it. Caching is also needed for advanced collaboration-features.
*Relevancy: ***** Implementation: Beta Phase*

Feed-Organization

The heart of the application: Adding and removing feeds, storing and arranging them inside of folders. The folders can be renamed and arranged as well, and will be represented in their own panel, with expand/collapse-options. An unread-count is attached to each feed and group.
*Relevancy: ***** Implementation: Beta Phase*

Frontpage-View

Similar to a newspaper-frontpage, the application will present the most relevant news and distinct recommendations by friends directly on the homepage of the application. The amount of displayed news will be customizable. In combination with an option to *mark all other news as read*, the frontpage provides a one-click overview of the most relevant information.
*Relevancy: ***** Implementation: Beta Phase*

---

10 *Scripting News Inc. 10 Feb. 2006 <http://www.opml.org/>*

Rating of Feeds and News-Items

Feeds can be prioritized by a rating of one through five stars, where five stars means "most relevant". The *five-star rating* is common to most users, e.g. from *Apple iTunes* – no further explanation on the rating-process is required. News-items on the other hand are automatically rated just by clicking on it. As news-items are mostly "throw-away"-content and must be processed in a large number, a distinct rating by the user is not effective and won't be accepted. The background-rating is unobtrusive and does not require any additional input from the user (conf. *3.4 Filtering Implementation*).
Relevancy: ***** *Implementation: Beta Phase*

Social Network

All users of the service will have a profile open to the public, with general information such as nickname, age, country, homepage and an avatar-picture. Each user has a personalized network-page, where he can search for other users by name or nickname – and add them as friends (conf. *3.5 Community Implementation*).
Relevancy: ***** *Implementation: Beta Phase*

Recommend News to a Friend

A *recommend-link* is attached to each news-item, which produces a popup with the user's current friend-list. He can mark some of all friends and send an annotated recommendation to them – a more direct form of *link-blogging*. Recipients will see these items on their start-page as well as highlighted in the river-of-news view. The targeted user has an option to subscribe to the originating feed if it's not already in his feed-list.
Relevancy: ***** *Implementation: Beta Phase*

Smart URLs

An often overlooked feature that greatly increases the usability of a site are *human-readable URLs*. They provide a breadcrumb-like look on the currently viewed site, e.g. *www.domain.com/feeds/frontpage*, are easy to remember and "hackable" by the user.
Relevancy: ***** *Implementation: Final Product*

Automatic Neighbor-Discovery

Aside from directly finding friends via a name-based search, the application could suggest new contacts by evaluating their reading behavior and matching feed-subscriptions.
Relevancy: ***** *Implementation: Final Product*

Automatic Feed-Discovery

In case the user does not know the URL of a specific feed, he can just point the application to the desired page and let it parse for any feed-information. This could also be helpful if a feed-URL is misspelled.
*Relevancy: ***** Implementation: Final Product*

Quarantine

A special folder is used to collect new feeds that are under "observation", so the user has time to evaluate the quality of the feed without "polluting" his normal feed-folders. After a specified amount of days, he might be reminded on the starting-page to decide about the future status of a quarantined feed.
*Relevancy: ***** Implementation: Final Product*

Public Feed-List

In social software, some info about other users should be available in a kind of public profile. In the case of an RSS-reader, this profile is basically the feeds a user reads. For privacy reasons, the user can disable public display of certain feeds.
*Relevancy: ***** Implementation: Final Product*

Automatic Rating by external Sources

Established services could contribute to a better relevancy-rating of news-items: So called *trackbacks* show interlinkage between blog-entries. Specialized blog-search-engines such as *Technorati.com* provide APIs to find popular news-items. Other factors such as Google-ranking or comment-counts can be an indicator for relevant feeds and posts.
*Relevancy: ***** Implementation: Final Product*

Feed-Statistics

In the mass of feeds a user can quickly loose overview on what he is subscribed to and how his subscriptions are actually relevant to him. A detailed statistical listing of each feed could help out: It could mark feeds that were inactive for a certain time-period as well as moved or deleted feeds. A ratio of read against unread news-items is computed as a long-term indicator of relevancy. The average rating by other users and their read-/unread-ratio put a specific feed into perspective. The overview-page will allow to unsubscribe from feeds or move them back into *Quarantine*.
*Relevancy: ***** Implementation: Final Product*

Help-System

In respective to the aimed user-group of power-users, the help-system will be held in a brief and problem-oriented FAQ-style. Additional help will be provided inline directly in the application.
*Relevancy: ***** Implementation: Final Product*

Localization

The application will be architected from the start to be localizable in other languages. For the beta-phase nevertheless, the only available language will be *English*, covering the largest user-base.
*Relevancy: ***** Implementation: Final Product*

History

Users might want to browse through old news which are already marked as read. A history-option allows to show all news back until a specified point in time. A date-ordered history of read news only could be also valuable to the user.
*Relevancy: ***** Implementation: Final Product*

Mobile Access

A filtered RSS-reader is also ideal for mobile access via cellphones and PDAs, where every byte costs real money and screen-size is limited. A user could for example quickly check the most relevant headlines on his starting page from his phone, delaying the rest of his news for a later browsing-session at a desktop-computer.
*Relevancy: ***** Implementation: Final Product*

Feature Overview

Alpha-Phase (due 2006-05-16)

Feature Title	Relevancy
Template-based Output	*****
RSS-/Atom-Viewing	*****
User-Management	*****
Bug Reporting	*****
OPML-Import/-Export	*****
River-of-News-View	*****

Beta-Phase (due 2006-06-01)

	Relevancy
Automatic Updates and Caching	*****
Feed Organization	*****
Frontpage-View	*****
Rating of Feeds and News-Items	*****
Social Network	*****
Recommend News to a Friend	*****

Final Product (further development)

	Relevancy
Smart URLs	*****
Automatic Neighbor Discovery	*****
Automatic Feed Discovery	*****
Quarantine	*****
Public Feed List	*****
Automatic Rating by external Sources	*****
Feed-Statistics	*****
Help-System	*****
Localization	*****
History	*****
Mobile Access	*****

3.4 Filtering Implementation

The whole product is based on one simple idea: *Provide a better signal-to-noise ratio for efficient news-reading.* Most of the analyzed competition with social features (conf. *2.3 Competitive Analysis*) get this point wrong: They provide additional information-clutter rather then reducing the existing data, e.g. through tagging and feed-recommendations. For these reasons, the first chapters will deal with features the product will *not* include for the sake of efficiency.

While this *discovery-approach* is working great in item-based scenarios such as music of movies, it is unnecessary for power-users who already have problems of keeping track with their current feeds. They are no beginners in the *blogosphere* and already have enough tools and resources where they could get new feeds.

Structuring feeds through *tags* and *smart folders* might sound logical at first, but in the end it just weakens the internal folder structure that is already existing: News are essentially *throw-away-items* much like a daily newspaper -- read once and then disposed. They aren't collected and archived, therefore tagging or rating them explicitly is just overkill.

In contrast, the presented application will focus on unobtrusive filtering-mechanisms, sometimes without even needing any specific user-actions: A clicked news-item will automatically gain a positive rating. A possible obstacle in this rating-behavior is the *cold-start problem*: A reliable rating is only gathered after a certain amount of persons read the news-item. To work around this problem, news will be pre-filled with a neutral rating, blended with the overall-rating of the feed. This blending-mechanism will also help to overcome the *fallover-effect*: news that already have a high rating by some users will be recommended to other users, becoming exponentially more popular in turn. The application has to algorithmically ensure an equal distribution of relevancy in this process, giving "average news" the chance to become popular as well.

Collaborative Filtering Theory

In the theory of collaborative filtering, two models for rating-predicition have established: Memory-based and Model-based. A memory-based scheme uses the similarity between two users to form a neighbourhood and predict a weighted average, e.g. *"User A and User B are neighbors by both liking News X, so when User A reads News Y it probably is interesting to User B as well"*. This method involves extensive analysis and works only if the data-set is sufficiently large. The Model-based scheme on the other hand analyzes on a item- rather than user-basis by building a item-relationship matrix, e.g. *"News X is similar to News Y in the matrix, so a User A who read News X might be interesting in News Y."* A special form of model-based prediction is the Slope One algorithm[11]: This method if far more effective than a linear analysis of all data-sets by computing an average difference between two items. Therefore it is most suitable for *non-cached realtime-querying* that is needed for news-analysis. It also weakens the *cold-start problem* because it does not need a complete neighbourhood to compute sensible predictions.

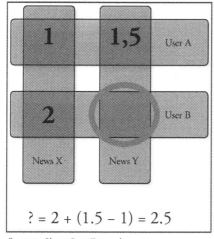

fig. 3-1: Slope One Example

3.5 Community Implementation

The social framework of the application will stay behind the primary goal: efficient news reading. It does not try to replicate community concepts such as *Friendster. com* or *OpenBC.com* with discussion-groups, extensive personal profiling and messaging-systems. While the interconnection with other user-ratings is crucial to the success of social news reading, community as such is just an utility. According-ly, the community features will restrict to adding and removing users on a personal friend-list. The only direct communication that is possible between users is through news-recommendations.

---

11 Lemire, Daniel and Anna Maclachlan. *"Slope One Predictors for Online Rating-Based Collaborative Filtering ".* 7 Feb 2005 <http://www.daniel-lemire.com/fr/documents/publications/lemiremaclachlan_ sdm05.pdf>

3.6 User Interface Design

Based on the promise of the application, the interface-design will be as efficient as possible and provide a rich experience to the user. Because it is an application-client rather than a website, the interface behavior will try to replicate typical desktop-metaphors such as foldable trees, context-menus or inline popups. Placement and wording of interface-elements will be oriented on established competition, to ease usage for possible switchers. Textual information other than the actual news-content will be reduced to a minimum, e.g. inline-help will be mainly substituted by an external FAQ-section. With the usage of modern web-techniques, a complete site-reload for simple tasks is not necessary, e.g. an email-address will be directly editable in place and save automatically.

The application has a mostly neutral color-space, with different shades of grey providing structure and occasional highlights in orange and yellow. The focus clearly lies on the provided content itself, and the interface tries to stay behind the scenes as much as possible without losing its structural importance. The application is architected for long scrolling pages, so important pieces of information have to recognizable in fractions of a second: e.g. headlines marking the beginning of a new feed are set in a large orange font, with visible spacing before and after the element. Additionally, the boundaries of each news-post are marked with a slightly darker background if the mouse is inside the active area, thereby signifying the change of content even if the user is scrolling fast without moving the cursor. If a newspost is clicked and by that marked as "read", it becomes semi-transparent to denote it's status (conf. *fig. 3-7 Four states of a newspost*). Secondary controls like the recommendation-form or feed-controls only show if the mouse is in the active area or has clicked an expansion-link - further removing unneeded clutter from the interface.

Typography is also optimized for readability, with a large "Lucida Grande"-typeface in both content and navigational areas. The content-area is restricted to a maximum width of 1000px, to stay in the specifications for optimal line-length of 80 to 120 characters[12]. Below this maximum the layout is *liquid*, adjusting to the browser's width and breaking overlapping elements into new paragraphs. The content-typography has a stronger contrast if the mouse hovers over it, again improving readability.

---

12 Ferrari , Tomás G., Carolina Short. *"Legibility and readability on the World Wide Web"* 2002
<http://bigital.com/files/Web_Legibility_Readability.pdf>

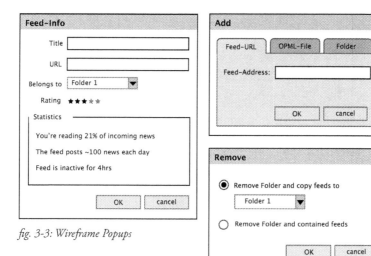

Feed-Info

Title

URL

Belongs to Folder 1

Rating ★★★☆☆

Statistics

You're reading 21% of incoming news

The feed posts ~100 news each day

Feed is inactive for 4hrs

OK   cancel

Add

Feed-URL   OPML-File   Folder

Feed-Address:

OK   cancel

Remove

◉ Remove Folder and copy feeds to
   Folder 1

○ Remove Folder and contained feeds

OK   cancel

fig. 3-3: *Wireframe Popups*

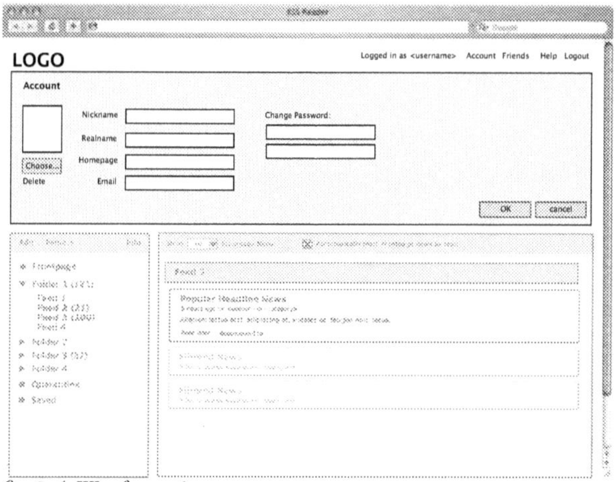

fig. 3-4: *Wireframe Account*

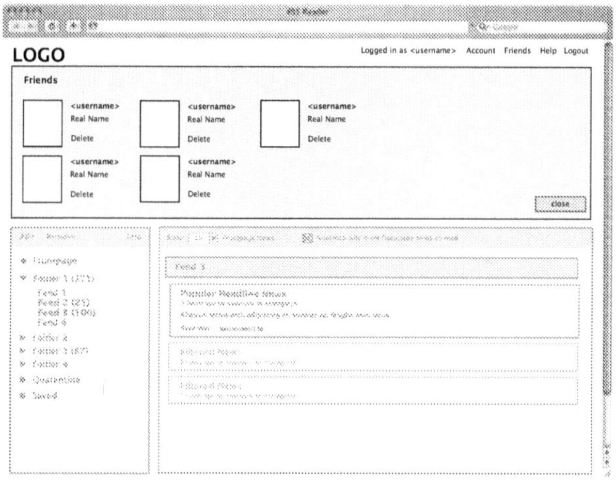

fig. 3-5: *Wireframe Friends*

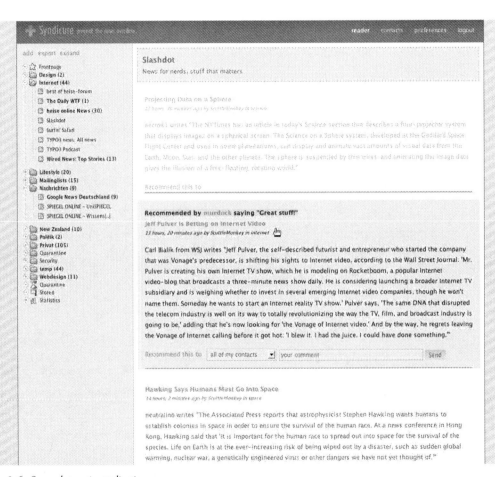

fig. 3-6: Screenshot main application

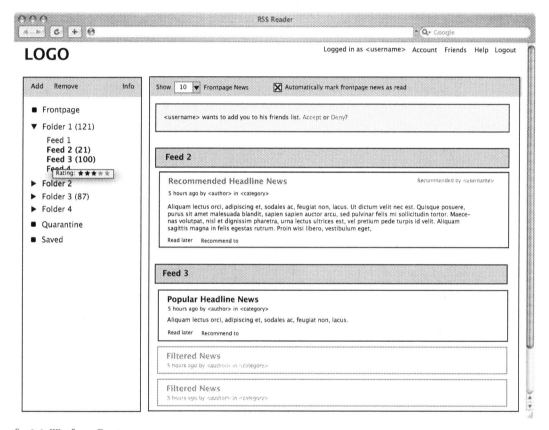

fig. 3-2: Wireframe Frontpage

fig. 3-7: Four states of a newspost (unread, unread-hover, read-hover, read)

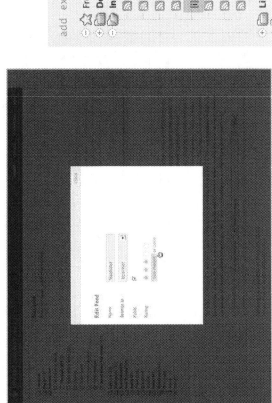

fig. 3-8: Inline-popups with darkened main window

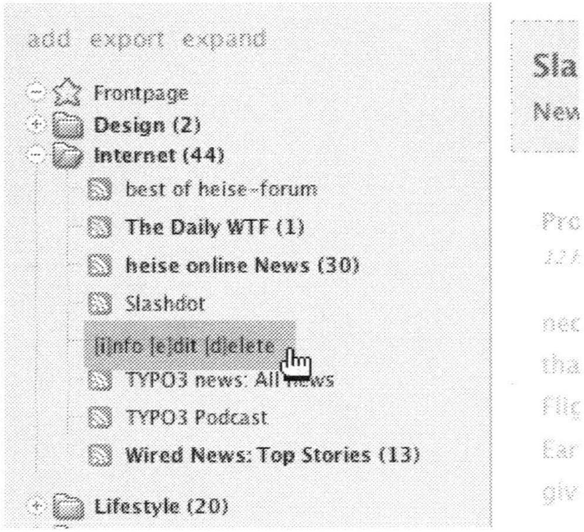

fig. 3-9: Context-Menu

3.7 Branding

A successful online application needs a distinct brand to stay in the attention of possible users. It has to convince quickly by both information and appearance. The most important building blocks are an easy-to-remember product name, a clear logo and probably a catchy subtitle. In optimal case, the combination of these three factors already conveys a basic understanding of the application. The field of online RSS-readers is already broadly populated, and the system needed a unique name that doesn't get confused with existent brands. The availability of a prominent TLD-domain (most likely in the *.com-space) is also an important criteria.

As the application mainly serves power-users who probably spend an unhealthy amount of hours on the internet, the idea of a *medical tool* quickly came up. Internet-surfing in general and RSS-reading in specific can turn into an addiction, the permanent urge to stay "up-to-date". The task of the application is to ease this addiction by pre-filtering news for the "patient". In the figurative sense, recommending news to each other is a global "self-help group" for users - a "virtual cure" from a real addiction. In combination with the unique word "*Syndication*", the process of publishing structured information for public usage, this leads to "*Syndicure*" - and the domain-name *www.syndicure.com*. The logo is enriched with a medical cross in orange, a broadly used color to indicate RSS-content. The subtitle reads "*prevent the news overdose*", another medical reference which also introduces the keyword "news".

fig. 3-10: Syndicure Logo

4. Technical Concept

4.1 Frontend

The graphical interface is produced entirely in W3C-compliant XHTML 1.0 Transitional[13]. Separation of style and content is achieved by CSS 2.0[14], while behavior such as user-feedback or popups is done with Javascript.

In the initial beta-phase the interface will be limited to a set of supported browsers (Safari 2.0+ and Firefox 1.0+) to keep efforts for testing and compatibility-issues as low as possible. Nevertheless, the system degrades gracefully on other browsers: If a non-supported platform is detected, all Javascript-behavior and advanced CSS-styling is turned off and the user gets a basic but functional system. By this separation of client- and server-behavior the system is still usable in low-end clients such as screenreaders or mobile phones (conf. *4.1 Mobile Access*).

XMLHttpRequest and DOM

To achieve the instant feedback of a web-application as described in *3.6 User Interface Design*, the system has to go beyond basic concepts of the web: A full page-reload on every user-interaction might work for content-based structures like normal web pages, but does not suffice for modern web-applications. Due to the HTTP-specification, all requests are directed from the client to a server ("pull"), while a server is not able to actively send unrequested data to a client ("push"). Therefore every enhancement of interaction has to happen on the client-side, with Javascript and partial HTTP-requests as supporting technologies. Every HTML-document is accessible to Javascript via the "*Document Object Model (DOM)*"[15]. The DOM is used to alter represented data without the need to poll a server, e.g. to provide a tree-like navigation that shows and hides subtrees without a page-refresh. By this dynamic feedback, an approximation to the GUIs of desktop-applications is possible.

The XMLHttpRequest-API is an essential building block for the "AJAX"-technique[16], which is used to implement partial page-requests in an already loaded HTML-document. This approach drastically reduces response-time in updating the presented data on the client or submitting new data to the server. In the current system, these technique is used to view feeds, add dynamic popups or save rating-values in the database. Because AJAX breaks certain browser-features such as history or progress-indication, a developer needs to ensure the usability of his application by manually implementing these features.

---

13 W3C Group IG, "XHTML 1.0: The Extensible HyperText Markup Language (Second Edition)". 1 Aug 2002 < http://www.w3.org/TR/xhtml1/>

14 W3C Group IG, "Cascading Style Sheets, level 2". 12 May 1998 <http://www.w3.org/TR/REC-CSS2/>

15 W3C Group IG, "W3 Document Object Model". 9 Jan 2005 <http://www.w3.org/DOM/>

16 Garret, Jesse James. „Ajax - a new Approach to Web Applications". 18 Feb 2005 <http://www.adaptivepath.com/publications/essays/archives/000385.php>

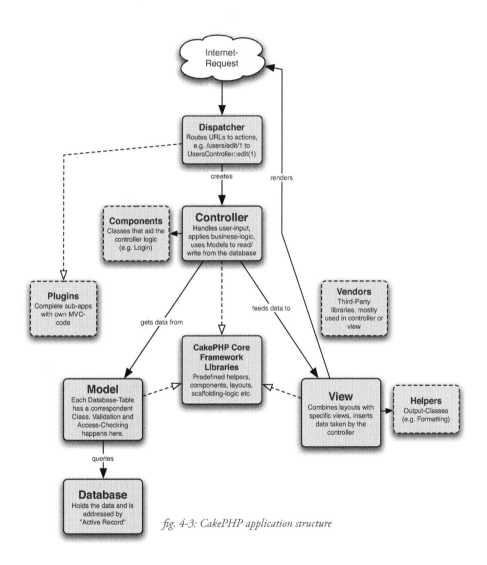

fig. 4-3: CakePHP application structure

Mobile Access

In most desktop-applications and traditional web-apps, producing a mobile version implied rewriting large parts of the codebase, greatly extending development- and maintenance-costs. With the advent of XHTML-compliant mobile browsers which are no longer bound to the WAP-protocol, producing a compatible application got a lot easier: In well-structured environments, it only needs some minor adjustments to style and behavior to adapt to reduced screen-space, navigational limitations and bandwidth-issues of mobile devices[17]. For the RSS-reader-application, a device-specific stylesheet displays smaller fonts and a more linear navigation, lacking javascript-interaction is compensated with a degradable system of server-requests. The application will automatically detect any mobile clients, and adjust the interface accordingly. In addition, the mobile-version is provided at
http://mobile.syndicure.com.

4.2 Backend

Business-Logic and database-structures are implemented in the popular open-source combination of PHP 4 and MySQL 4. This broad support enables the choice between various third-party modules and frameworks, It also enables a rapid prototyping process that is possible with the instant feedback of an interpreted language.

The MVC-Pattern

The "Model-View-Controller"-paradigm is a widely accepted "software design pattern"[18]. It states that an application should be seperated into three parts: data model, user interface, and control logic — to provide enhanced modularization, maintainability and reusability.

"The essential purpose of MVC is to bridge the gap between the human user's mental model and the digital model that exists in the computer. The ideal MVC solution supports the user illusion of seeing and manipulating the domain information directly. The structure is useful if the user needs to see the same model element simultaneously in different contexts and/or from different viewpoints."[19]

Trygve Reenskaug, Developer of Smalltalk and inventor of MVC

fig. 4-1: Syndicure Mobile

17 Opera Software ASA. "The Phone Factor - My Opera Community" 2006 <http://my.opera.com/community/dev/mini/phones/>
18 Gamma, Erich, R. Helm, R. Johnson, and J. Vlissides. "Design Patterns", Addison-Wesley, 1995
19 Reenskaug, Trygve M. H.. "MVC XEROX PARC 1978-79". 25 May

Model: Represents the data persistence layer (mostly a database). As the model does not have any knowledge of it's control logic or representation to the user, it does not maintain any application-state. Validation and basic data-specific tasks are extended responsibilities of a model in a web-application.

View: Renders the model into a suitable form of interaction, mostly a HTML-document with links and form-controls. Controllers which are connected to views are triggered via HTTP-requests.

Controller: This intermediate instance between view and model responds to user-actions and triggers data-changes. Advanced data-handling and business logic such as authentication happens here.

Especially in complex web-applications the MVC-approach has many advantages: Frequently changing frontend-templates can be altered and substituted safely without affecting the underlying data-integrity. In contrast to the popular mix of HTML and scripting languages in one document, a separated template is accessible even to team-members without programming experience such as interface-designers. MVC also facilitates a clean security-model, by keeping application-state and management of HTTP-requests in the controller, while the view is the only accessible entity that is rendered to the client.

The CakePHP Framework

Thoroughly applying an MVC-pattern from scratch is a tedious task, and due to the time-restrictions of the implementation-phase the choice for a pre-existing solution was obvious. CakePHP[20] is a so called "rapid development framework", a term coined by IBM in the 1990s[21] and revived in recent times by the framework "Ruby on Rails"[22]. "Rapid development" strives to increase speed and quality in the development process by adhering to generally acknowledged specifications – the mantras are "convention over configuration" and "don't repeat yourself". CakePHP tailored this approach to web-specific needs, such as:

- Object-oriented programming
- Database-abstraction layer
- Object-relational mapping ("Active Record")
- Scaffolding (automatic interface-generation)
- Validation, data-sanitization and automatic URL-to-action mapping
- Components for security-, session and cache-handling
- Access Control Lists

The concept of "*Object Relational Mapping*" copes with the discrepancy between data-storage in relational databases and object-oriented programming: Database-tables map to classes, rows in a table map to objects of the class, and database-columns are represented by object-attributes. "*Active Record*"[23] extends this concept by supplying the developer with a set of functions for getting and setting data, while

2006
<http://heim.ifi.uio.no/~trygver/themes/mvc/mvc-index.html>
20 Cake Software Foundation. "CakePHP: the rapid development php framework". 25 May 2006
<http://www.cakephp.org>
21 Martin, James. "Rapid Application Development". May 1991. Macmillan Col Div.
22 37 Signals. "Ruby on Rails". 25 May 2006 <http://www.rubyonrails.org>
23 Martin Fowler. "P of EAA: Active Record". 25 May 2006 < http://www.martinfowler.com/eaaCatalog/ activeRecord.html>

minimizing the configuration-overhead. Adding, removing, and changing attributes of Active Record objects instantly reflects in the database. By semi-automatic detection of table-relations, complex queries across multiple tables are simplified - direct SQL-queries are barely needed due to the wrapper-functionality.

The developer in a rapid development framework has to study conventions first, and profits greatly from implementing them accordingly. Especially in terms of *validation* and *data-sanitization*, CakePHP builds up a framework that is easily comprehensible and extendable: Through numerous callbacks in the model- and controller-classes, he is able to transparently inject custom behavior, such as an additional access-check after querying the database or switching to a mobile layout before rendering.

Scaffolding describes a crucial process to rapid development: The ability to modify and control the data without the need to set up a GUI. Based on the automatically detected data-structure, the views to create, edit and delete records are generated by the application itself. *Scaffolding* is mostly used in early development-stages to sketch out functionality, as it lacks the flexibility to cope with advanced interface design.

CakePHP also incorporates an extended set of "classic" framework-features, modularized in *Helpers* and *Components*. These classes help to accomplish common tasks like authentication, session-management or text formatting. Developers are encouraged to write custom modules where functionality can be clearly separated and reused in other parts of the application.

Entity Relationship Model

The application basically manages users, their information (feeds and feeditems), their relations to other users and item-specific meta-data (such as statistics). To maintain security in such a multi-user environment, some entities have to be divided into an abstract system-wide entity and a user-specific entity (*fig. 4.3: Application Entity Relationship Model*).

Users: Saves user-information, md5-hashed passwords and user-specific application-preferences. A photo-file can be referenced via the *avatar*-column.

Feeds/Userfeeds: To maintain an unique relationship between users and their subscribed feeds, each feed is saved as a system-wide item and a user-specific item. A userfeed does not duplicate information that is already present in the feed, but adds overwrite-columns such as a custom title. By this separation, a user is able to unsubscribe from a userfeed without actually deleting the underlying feed - even if he was the only subscriber, the feed is kept for statistical reference.

Feeditems/Userfeeditems: Store single news-items from a feed, and are also divided in system- and user-specific items. Hashes are created from the supplied meta-data to uniquely identify an item: An *unique_hash* that shouldn't change through the lifetime of a news-item is calculated from creation-timestamp and title. An *update_hash* indexes the content of each item, to notify the user of any changes and eventually re-display an already viewed feeditem. The *update_hash* is stored both in a userfeeditem and a feeditem to enable to enable comparison of „versions". Userfeeditems are only created for a specific user when he actually views the feeditem in the overview. If the user clicks on a link inside a feeditem, it is marked as „read" (*read_at*). This timestamp doubles as a statistical signature for the read-status of a specific item.

Relationships: Describes a user's friends-list. As relationships don't have to be mutual, they are saved with a direction instead of an n:m relationship (*user_id* and *foreignuser_id*).

Acos/Aros/Aros_acos: Some database-items (Users, Userfeeds, Folders) are managed by „Access Control Lists", some remain without access-control to maintain performance. The managed items are not directly related via foreign-keys, but bear a custom unique alias (e.g. a userfeed with id=99 is stored as the string "User.99"). This approach makes ACL-checks expandable to entities not stored in a database, such as controller actions. (*conf. 4.2.5 Multi-user access checking*)

Slope_one_dev: Caching-table to quickly access results of the „nearest neighbour algorithm" (*conf. 3.4 Filtering Implementation*).

Feed_statistics/Feed_monthly_statistics/Feeditem_statistics: Tables with partly redundant information to store statistical data. The (mostly numbered) results are cached in this table rather than querying the real data at each access. In addition, the may have a longer lifespan than the actual data (e.g. year-averages on feeditems where the actual feeditems are already removed from the generating source).

Class Diagram

All implemented classes revolve around the MVC-concept outlined by CakePHP and extend basic framework-classes. Figure *4.5 Class Diagram* outlines only custom classes without the inherited framework-basis. It also shows logic connections of classes rather than strict inheritage, and groups them into model, view and controller to better resemble the application-structure. Accordingly, only relevant attributes and functions are included.

Most of the inner workings can be found in the custom controllers. Not all controllers have view and by that are not supposed to be user-accessed - they provide low-level functionality that is normally executed by the administrator or a commandline-task (cronjob). Some models are managed by a common controller, e.g. adding of Folders and Userfeeds share a common interface in the *ReaderController*.

FeedUpdater is not directly triggered by the User, and therefore has no views.

ReaderController has a central interface for adding Userfeeds, Folders and importing OPML, therefore the corresponding views do not exist in the specific controllers. It also provides most of the frontend-functionality, such as displaying feeds and the subscription-tree. Although the controller implements the methods *view_feed()* and *view_folder()*, they are both rendered by a common view.

Feed/Feeditem are only indirectly altered by the user by adding a Userfeed, therefore they have no distinct controller or view.

ACM ("Access Control Lists Manager") is a third-party plugin that is only accessible to the administrator.

Location	Description
index.php	Dispatcher, instanciates controllers based on the URL
cake/	Core libraries (not altered by the developer)
cakelibs/	Important core classes
cake/libs/controller/components/	Core classes aiding the controller logic
cake/libs/view/helpers/	Core classes for modifying output in views
app/	Application-specific code
app/config/	Configuration of database-access, global constants, URL routing
app/app_controller.php	Base-controller with authentification and access-checking
app/controllers/	Custom controllers (most business-logic resides here)
app/controllers/cli/	Command-line scripts triggered by cron (not accessible to the user)
app/controllers/components/	Custom classes aiding the controller logic
app/app_model.php	Base-model with validation-rules and access-checks
app/models/	Custom Models
app/views/	View related files
app/views/layouts/	Global page-layouts
app/views/errors/	Custom error messages
app/views/helpers/	Custom classes for modifying output in views
app/webroot/	Public accessible folder with static content
app/webroot/css/	Cascading Stylesheets
app/webroot/img/	Images for public display
app/webroot/js/	Javascript Behaviour
vendors/	Third-party libraries
tmp/	Used for logs, view caching and query caching

fig. 4-2: CakePHP directory and file structure

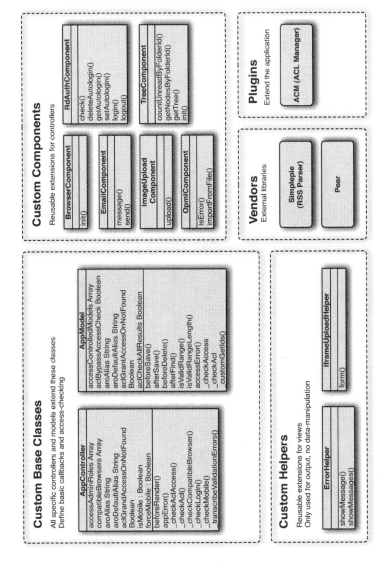

Custom Base Classes

All specific controllers and models extend these classes
Define basic callbacks and access-checking

AppController
accessAdminRoles Array
compatibleBrowsers Array
aroAlias String
aroDefaultAlias String
aclGrandAccessOnNotFound Boolean
isMobile : Boolean
forceMobile : Boolean
beforeRender()
appError()
_checkAclAccess()
_checkAcl()
_checkCompatibleBrowser()
_checkLogin()
_checkMobile()
_transcribeValidationErrors()

AppModel
accessControlledModels Array
aclBypassAccessCheck Boolean
aroAlias String
aroDefaultAlias String
aclGrandAccessOnNotFound Boolean
aclCheckAllResults Boolean
beforeSave()
afterSave()
beforeDelete()
afterFind()
isValidRange()
isValidRangeLength()
accessError()
_checkAccess
_checkAcl
_customGetIds()

Custom Components

Reusable extensions for controllers

BrowserComponent
init()

EmailComponent
message()
send()

ImageUpload Component
upload()

OpmlComponent
isError()
importFromFile()

RdAuthComponent
check()
deleteAutologin()
getAutologin()
setAutologin()
login()
logout()

TreeComponent
countUnreadByFolderId()
getNodesByFolderId()
getTree()
init()

Custom Helpers

Reusable extensions for views
Only used for output, no data-manipulation

ErrorHelper
showMessage()
showMessages()

IframeUploadHelper
form()

Vendors

External libraries

Simplepie (RSS Parser)

Pear

Plugins

Extend the application

ACM (ACL Manager)

fig. 4-3: Class diagram

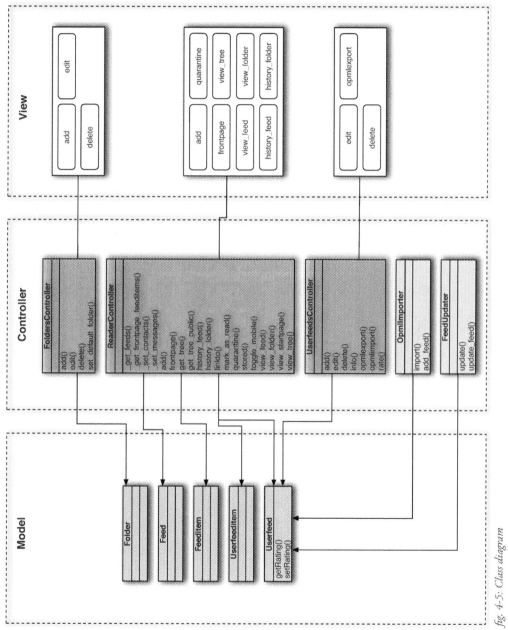

fig. 4-5: Class diagram

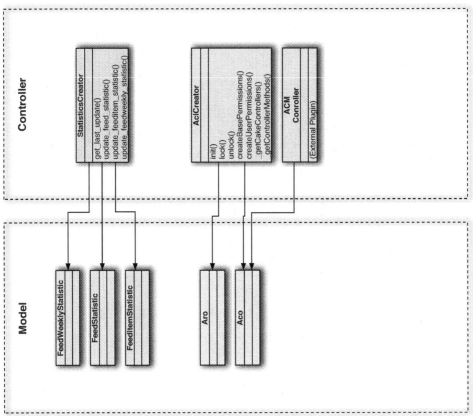

fig. 4-6: Class diagram

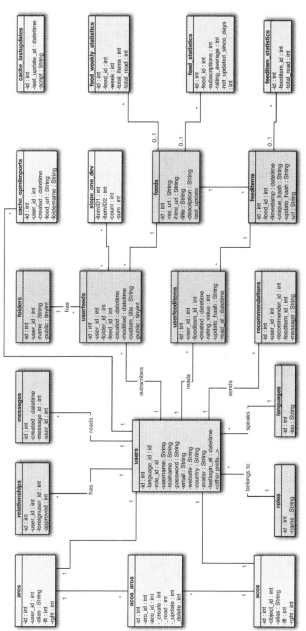

fig. 4-7: Application Entity Relationship Model

49

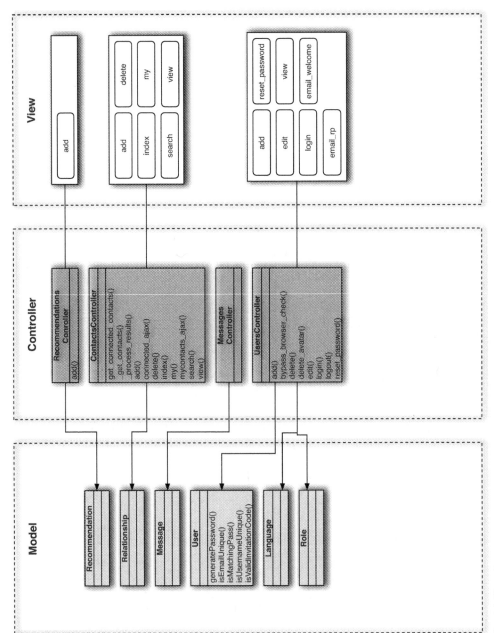

fig. 4-8 Class diagram

50

Mutli-user access control

Access-control in a multi-user environment is a two-step process: An application has to provide a form of authentication and role-management. But it also has to manage item-user relations and control access to these items. Authentication and access control is implemented in the controller-logic, mostly by some checks directly in the respective methods, e.g. if the user-id in an edit-request matches with the user-id stored in the session. Every check relies on a persistence layer where user-credentials are saved, mostly PHP-sessions that connect a HTTP-request to a specific user.

All direct user-authentification such as login/logout-methods happen in *UsersController*, while sessions are managed by *RdAuthComponent* and the built-in *SessionComponent*. User-entities are currently categorized into the following rolse: *Everybody* (not logged in), *User* (standard behavior) and *Administrator* (full access).

Unfortunately, method-based access-checks are redundant and error-prone, as there are no globally defined restrictions. CakePHP provides an alternate, more manageable and robust approach: *Access Control Lists* (ACL)[24]. This security-concept relies on abtraction and privilege separation, allowing a fine-grained control. In ACL-terms, users are *Access Request Objects (ARO)*, who in turn request *Access Control Objects (ACO)*. Both AROs and ACOs are identified with an unique alias and can be infinitely nested. By this aliasing, ACLs are not restricted to a certain authentication- or content-structure - they can contain everything from URLs over database-rows to single object properties. Each connection between an ARO and ACO signifies a permission, which can be either *allow* or *deny*. Optionally, these permission can be set seperately for a common action, such as *create, read, update* and *delete*. Permissions are calculated in a cascade where the most specific permission to an item applies (e.g. the group "user" is allowed to *read* other users and denied everything else, but the owner of a record has another relationship which also allows *edit* and *delete* actions). See *fig. 4.7 ACL example* for a more detailed explanation.

In CakePHP, AROs essentially represent groups and users, whereas ACOs are either controller actions or model-items. Access on these objects is checked on a application-wide basis in both model and controller, by injecting callback-methods into the model-controller communication.

---

24 *Wikipedia. "Access Control Lists". 26 May 2006 < http://en.wikipedia.org/wiki/Access_control_lists>*

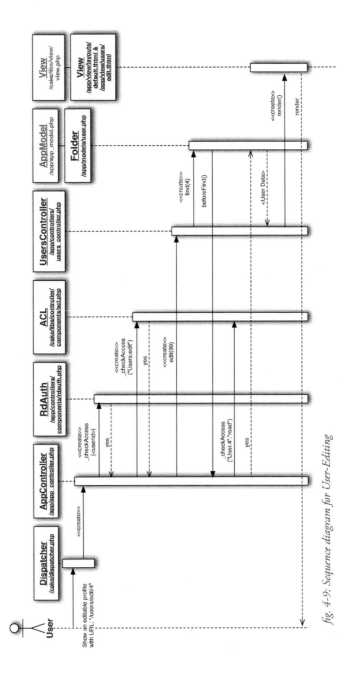

fig. 4-9: Sequence diagram for User-Editing

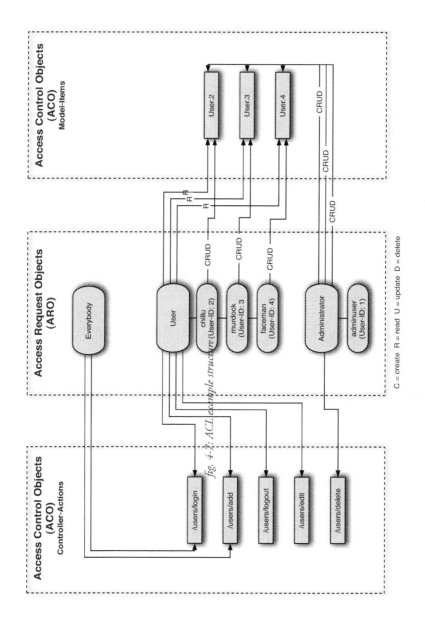

fig. 4.7: ACL example structure

5. Outlook

The concept of web-based software and smart information-aggregation with RSS is rather new, even by internet-standards. As the world of online-semantics is evolving, the implemented prototype also leaves plenty of room for progress. Due to it's open-source licensing model, chances are good that other users will discover the advantages of smart RSS-reading and join further development.

The main strength of the application is collaboration between it's users, by recommending and sharing news. The practical implications and success of this approach still have to be proven - the prototype was only used by early beta-testers. A "critical mass" of users with similiar interests and a shared social network is required to actually add value to the single user. This mass will be achieved by actively promoting *Syndicure*, through the means described in *3.1. Marketing*. During this process, public beta-testing will ensure the quality and usability of the product - an important analytical step that was not taken in this concept due to time-restrictions.

Syndicure already is a full-fledged RSS-reader, with most basic and some advanced features in place and ready for daily usage. Nevertheless it can be enhanced: New users will be comforted with a brief help-system and introductory wizards to import and rate existing feeds or subscribe to examplary collections. The rating-algorithm could be improved by supporting external rating-sources such as Amphetarate[25] or Technorati[26]. On a technical level, performance-improvements and scaling-plans are cruicial to support a larger user-base - at the moment a dedicated server might have problems handling more than several dozens of active users.

The system demonstrates how much can be achieved in web-based software-development with very little ressources. Through the implementation of open-source software, it gets easy to develop on a minimal budget, timeframe and manpower - and nevertheless become a small and agile competitor for large coorporations. Syndicure is one of many examples proving that most functions of monolithic desktop-software are also usable as „web-based software". It is a small step towards the shift from classic workstations with precompiled software to a „web-based operating system" that only needs a browser and internet-connectivity on the client-side.

The developed system and it's underlying concept also demonstrate the need for smart and agile information-tools to cope with the massive amount of online-data. It uncovers the missing semantics of most information published online - standardized relationships have yet to be established to make the step from simply storing letters to intelligently relating pieces of information. Syndicure can only serve as a preliminary and isolated tool to fulfill the current demand - solving this problem poses a more global task.

---

25   Unknown. "Sourceforge.net: Amphetarate RSS Recommendation Server". 12 Jun 2006 < http://
sourceforge.net/projects/amphetarate/>
26   Technorati Inc. "Technorati" 12 Jun 2006 <http://technorati.com/>

Printed Ressources

1. 37signals, Matthew Lindermann and Jason Fried. "Defensive Design for the Web". 2 Mar. 2004. New Riders Press.
2. Burkard, Remo. "Synergies between Information and Knowledge Visualization". 2005. Springer Lecture Notes in Computer Science
3. Gamma, Erich, Richard Helm, Ralph Johnson, and John Vlissides. "Design Patterns", Addison-Wesley, 1995
4. Martin, James. "Rapid Application Development". May 1991. Macmillan Col Div.
5. Maeda, John. "Creative Code". 2004. Thames & Hudson Inc.

Online Ressources

1. 37 Signals. "Ruby on Rails". 25 May 2006 <http://www.rubyonrails.org>
2. Advanced Insights. "New Marketing via social software". 14 Sept. 2005. 10 Feb. 2006.
   <http://www.advancinginsights.com/mybiz/new_marketing_via_social_software>
3. Bradford Paley, W. "Textarc.org". 10 Feb. 2006 <http://www.textarc.org>
4. Breese, J., D. Heckerman and C. Kadie. . "Empirical analysis of predictive algorithms for collaborative filtering". July 1998. 10 Feb. 2006. <ftp://ftp.research.microsoft.com/pub/tr/tr-98-12.pdf>
5. Cake Software Foundation. "CakePHP: the rapid development php framework". 25 May 2006 <http://www.cakephp.org>
6. Ferrari , Tomás G., Carolina Short. "Legibility and readability on the World Wide Web" 2002
   <http://bigital.com/files/Web_Legibility_Readability.pdf>
7. Fowler, Martin. "P of EAA: Active Record". 25 May 2006 <http://www.martinfowler.com/eaaCatalog/activeRecord.html>
8. Garret, Jesse James . „Ajax - a new Approach to Web Applications", 18 Feb 2005
   <http://www.adaptivepath.com/publications/essays/archives/000385.php>
9. Heylighen, F. "Collaborative Filtering". 31 Jan. 2001. 10 Feb. 2006. <http://pespmc1.vub.ac.be/COLLFILT.html>
10. Sarwar, B., G. Karypis, J. Konstan and J. Riedl. . "Item-based Collaborative Filtering Recommendation Algorithms", 19 Feb. 2001. 10 Feb. 2006. <http://www10.org/cdrom/papers/519/>
11. Hinchcliffe, Dion. "Notes on making good social software". 2005. 10 Feb. 2006.
    <http://web2.wsj2.com/notes_on_making_good_social_software.htm>
12. Kleinberg, Jon, and Mark Sandler. "Using Mixture Models for Collaborative Filtering". 2004. 10 Feb. 2006.
    <http://www.cs.cornell.edu/home/kleinber/stoc04-mixture.pdf>
13. Konstan, Joseph. "Recommender Systems: User Experience and System Issues". 3 Dec. 2000. 10 Feb. 2006.
    <http://www.grouplens.org/papers/JK-GL-Summer-2005.pdf>
14. Lemire, Daniel, Sean McGrath. "Implementing a Rating-Based Item-to-Item Recommender System in PHP/SQL". Jan. 2005. 10 Feb. 2006. <http://www.daniel-lemire.com/fr/documents/publications/webpaper.pdf>
15. Lemire, Daniel, and Anna Maclachlan. "Slope One Predictors for Online Rating-Based Collaborative Filtering ". 7 Feb 2005. 10 Feb. 2006. <http://www.daniel-lemire.com/fr/documents/publications/lemiremaclachlan_sdm05.pdf>
16. Opera Software ASA. "The Phone Factor - My Opera Community" 2006
    <http://my.opera.com/community/dev/mini/phones/>
17. Reenskaug, Trygve M. H.. "MVC XEROX PARC 1978-79". 25 May 2006
    <http://heim.ifi.uio.no/~trygver/themes/mvc/mvc-index.html>
18. Resnick, Paul, and Neophytos Iacovouet. "GroupLens: An Open Architecture for Collaborative Filtering of Netnews". 1994. 10 Feb. 2006. <http://www.si.umich.edu/~presnick/papers/cscw94/GroupLens.htm>
19. Röthlingshöfer, Bernd. "Manifesto zur Kunden-Evangelisation". 6 Dec. 2004. 10 Feb. 2006.
    <http://berndroethlingshoefer.typepad.com/smc/2004/12/manifesto_zur_k.html>
20. Technorati Inc. "Technorati" 12 Jun 2006 <http://technorati.com/>
21. Spolsky, Joel. "Painless Software Schedules". 29 Mar. 2000. 10 Feb. 2006.
    <http://www.joelonsoftware.com/articles/fog0000000245.html>
22. Unknown. "Sourceforge.net: Amphetarate RSS Recommendation Server". 12 Jun 2006
    <http://sourceforge.net/projects/amphetarate/>
23. W3C Group IG, "W3 Document Object Model". 9 Jan 2005 <http://www.w3.org/DOM/>
24. W3C Group IG. "Cascading Style Sheets, level 2". 12 May 1998 <http://www.w3.org/TR/REC-CSS2/>

25. W3C Group IG, "XHTML 1.0: The Extensible HyperText Markup Language (Second Edition)". 1 Aug 2002 < http://www.w3.org/TR/xhtml1/>

26. Wikipedia. "Access Control Lists". 26 May 2006 <http://en.wikipedia.org/wiki/Access_control_lists>

27. Wikipedia. "Collaborative Filtering". 10 Feb. 2006. <http://en.wikipedia.org/wiki/Collaborative_filtering>

28. Wikipedia. "Open-Source Funding". 13 Feb. 2006 <http://en.wikipedia.org/wiki/Open_source_funding>

29. Wikipedia. "Slope One". 10 Feb. 2006. <http://en.wikipedia.org/wiki/Slope_One>

Figures

Online-Demo

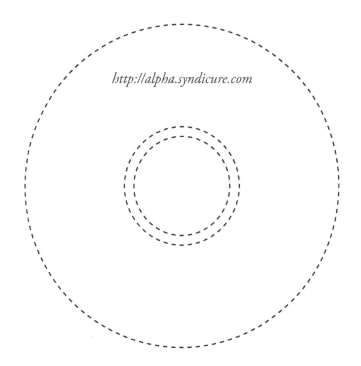

http://alpha.syndicure.com

PDF

http://www.chillu.com/docs/thesis_collaborative-filtering-with-rss.pdf

Thesis Presentation

http://www.chillu.com/docs/presentation_collaborative-filtering-with-rss.pdf

www.ingramcontent.com/pod-product-compliance
Lightning Source LLC
LaVergne TN
LVHW080104070326
832902LV00014B/2416